# GEOMETRY

## FOR KIDS IN 3D

### A-SELF TEACHING GUIDE

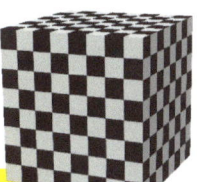

PARRELLOGRAM

HEXAGON

RIGHT ANGLE TRIANGLE

TRAPEZOID

# Preface

This book is for children who want to learn the basic concepts of basic geometry. For those who which to use it for home schooling children it can be very useful. The shapes are introduced with more details on each shape. This can help if your child in in middle school there is a short quiz at the end the book.

# Acknowledgement

Illustrations in this book generated using Microsoft word. Adobe illustrator.

Copyright 2018 by Joseph Thompson All rights reserved. Created in the United States of America.

This page was left blank intentionally

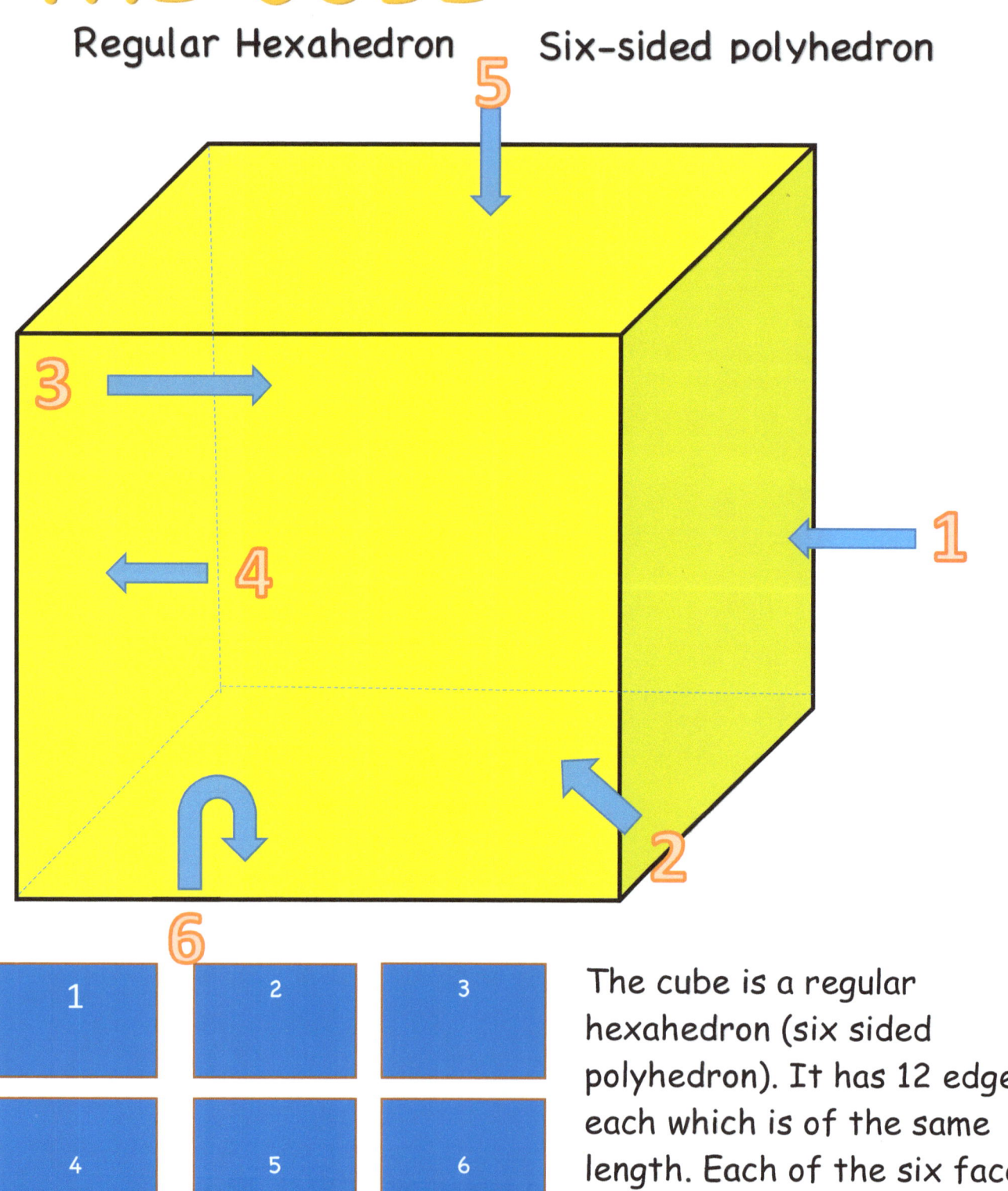

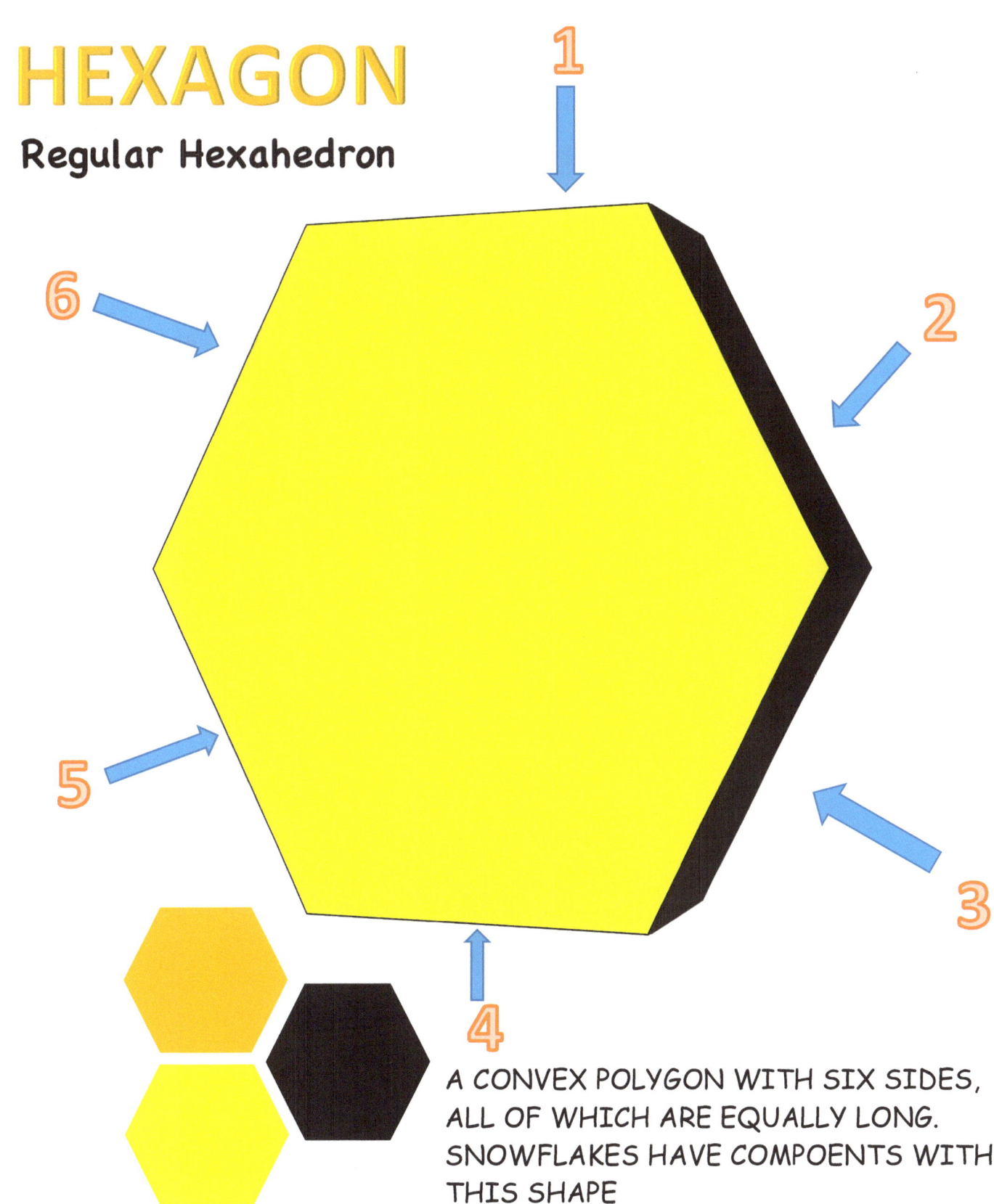

# CIRCLE

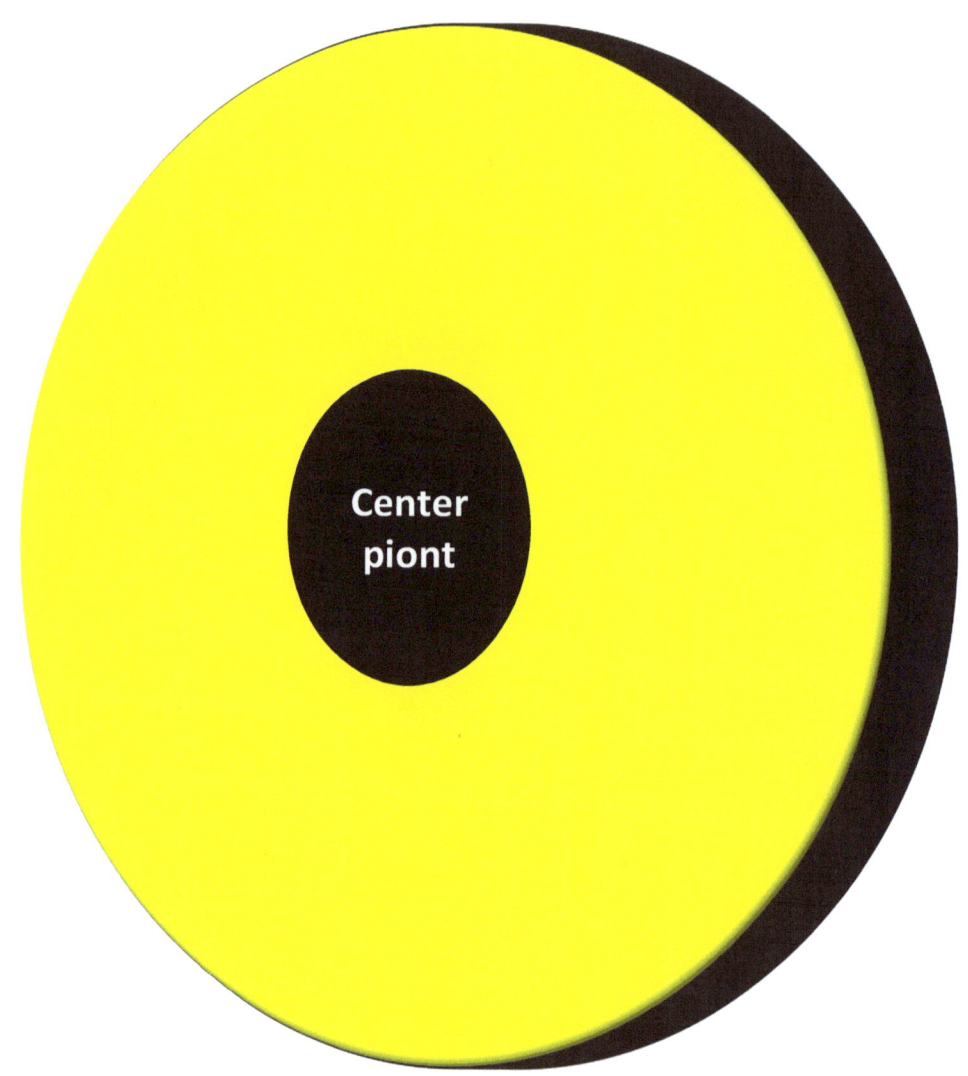

A circle is a geometric figure consisting of all points in a plane that are equally distant from some center point. The circle is an example of conic sections. This term came from the fact the circle can be defined as sets of points resulting from the intersection of a plane with a cone

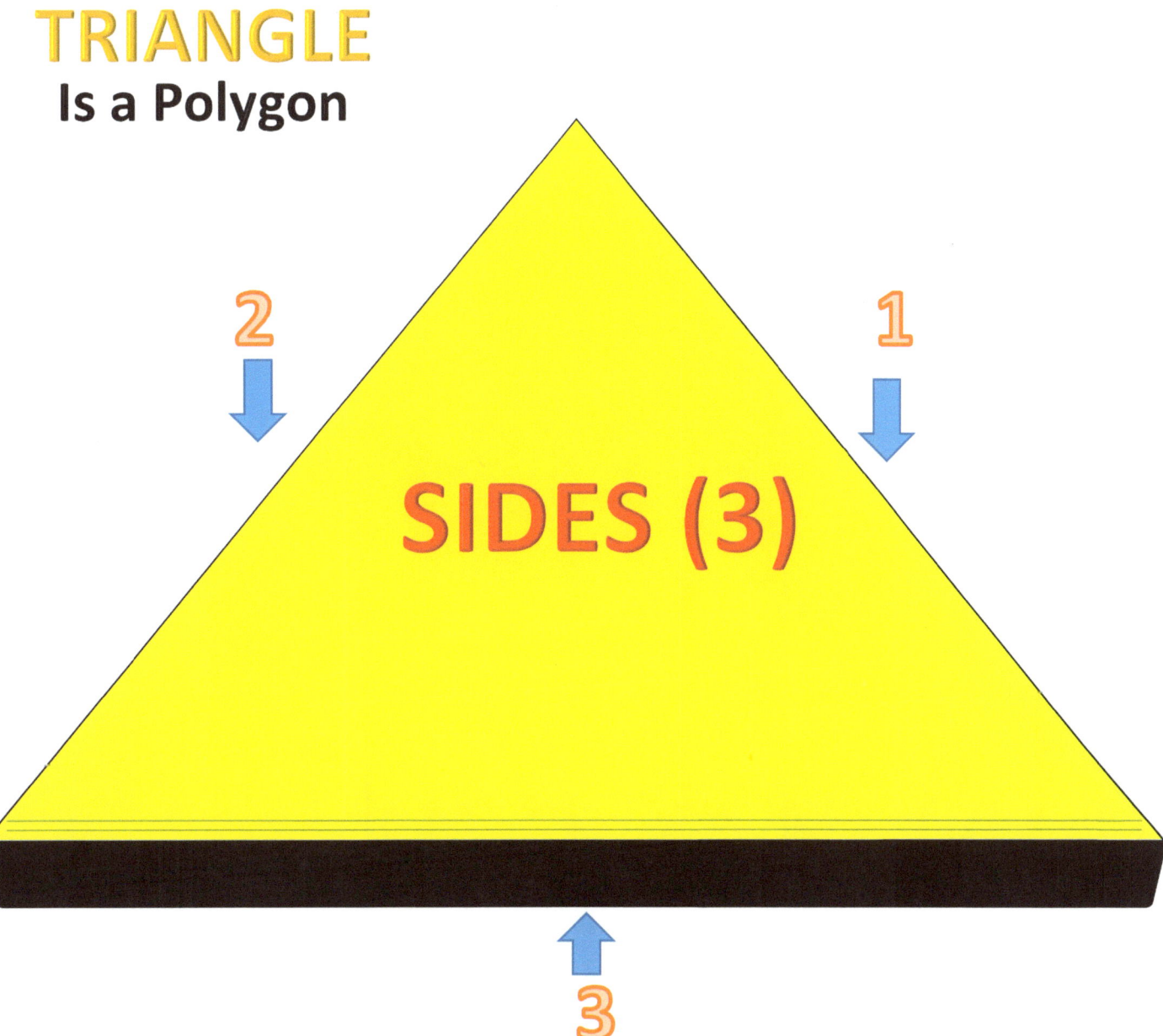

TRIANGLES ARE POLYGONS WITH THE LEAST POSSIBLE NUMBER OF SIDES (THREE)

# CYLINDER

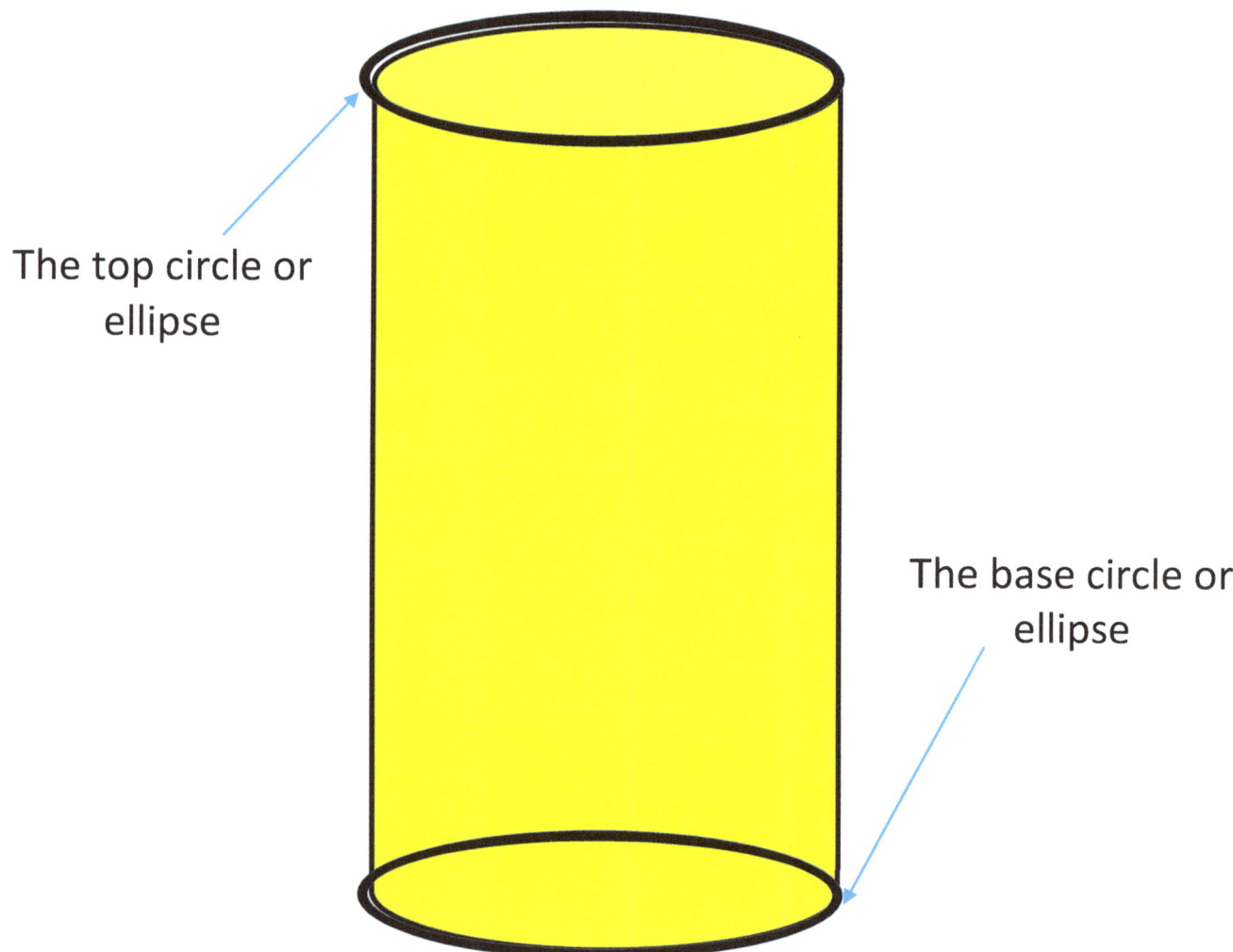

The top circle or ellipse

The base circle or ellipse

A cylinder has a circular or elliptical base, and a circular or elliptical top that congruent to the base and that lies in plane parallel to the base. A cylinder itself consists of the union

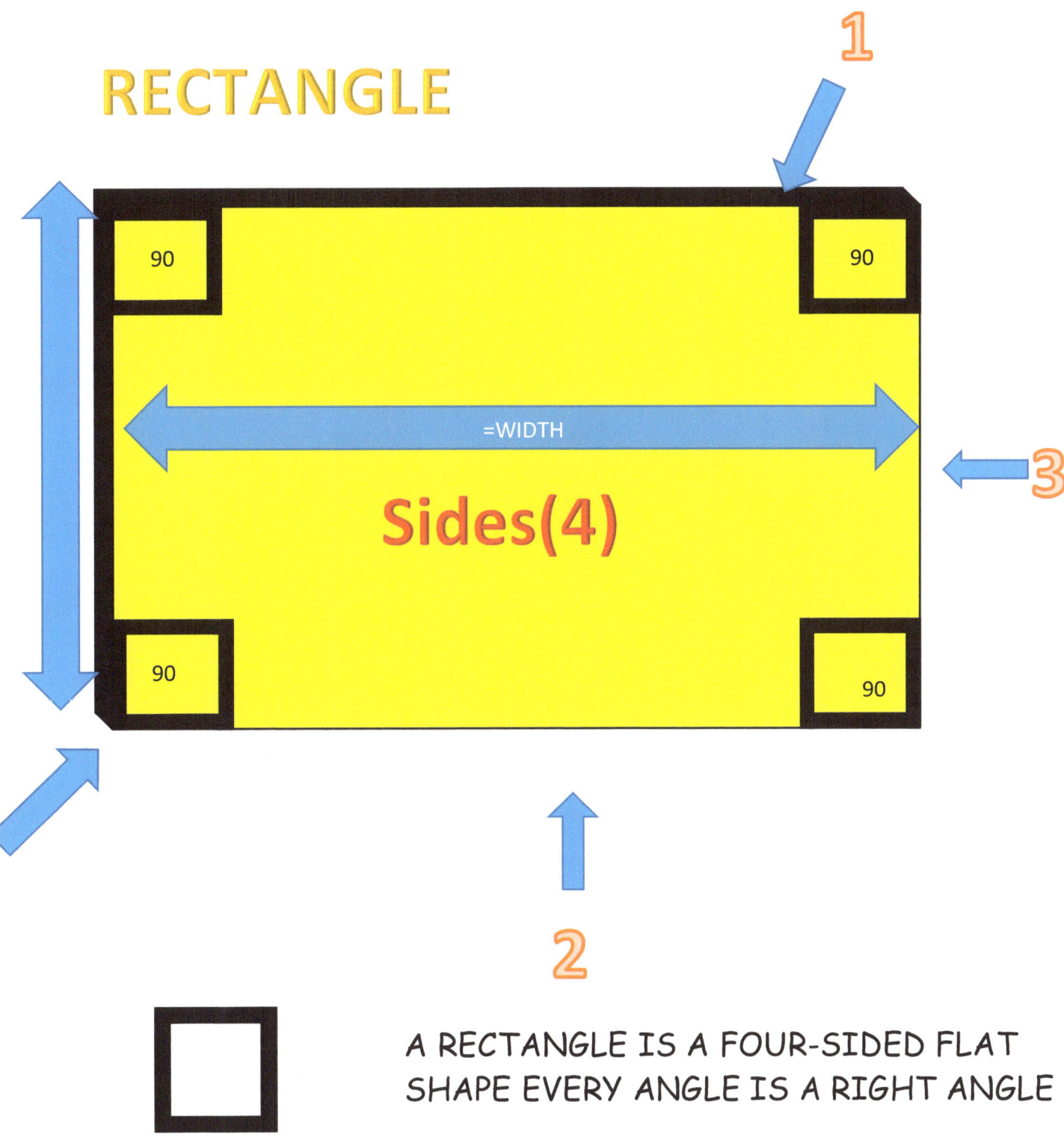

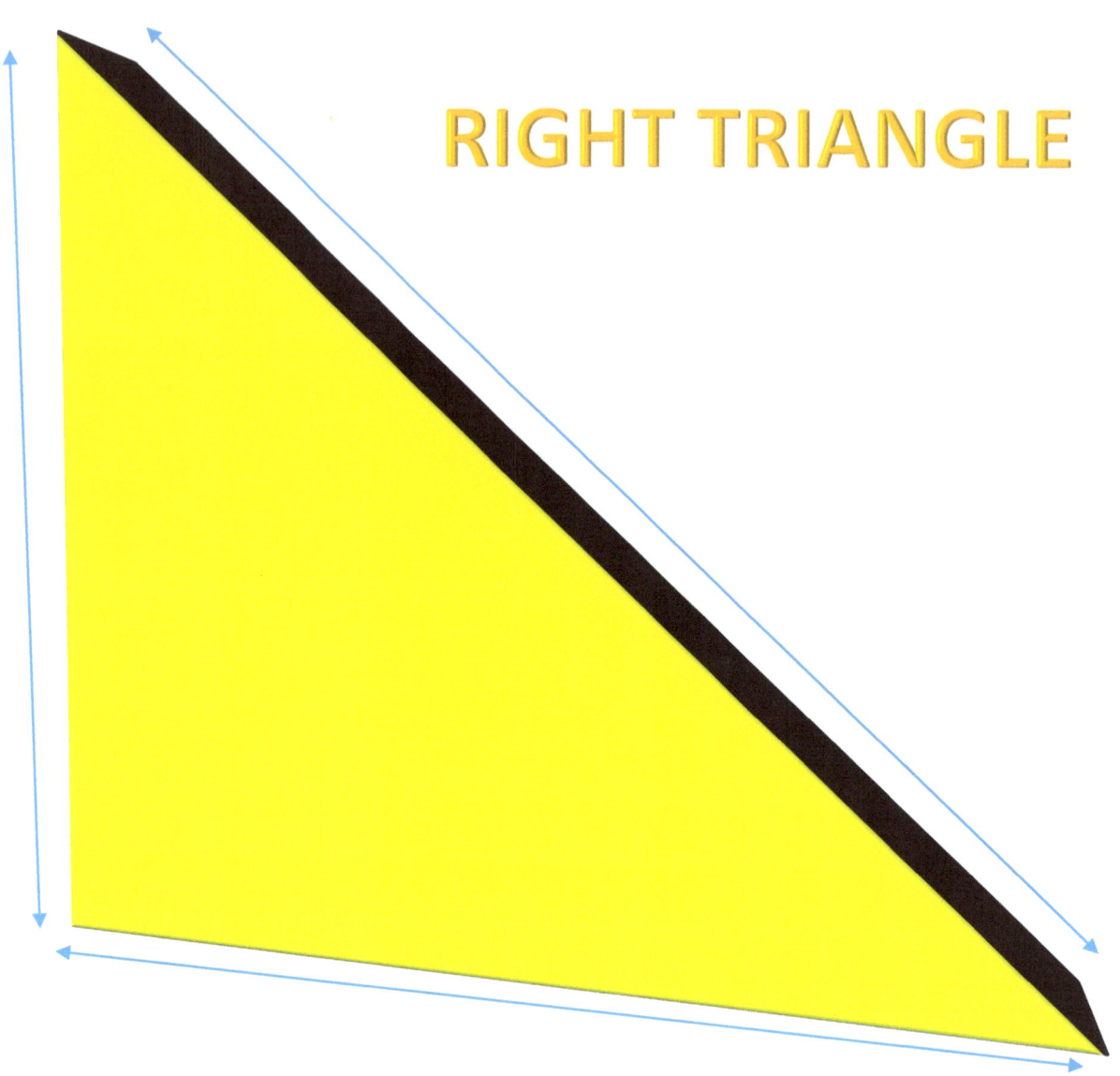

# RIGHT TRIANGLE

A right triangle is a triangle in which one angle is a right angle. The relation between the sides and angles of a right triangle is the foundation for trigonometry. The side opposite the right angle is known as the hypotenuse.

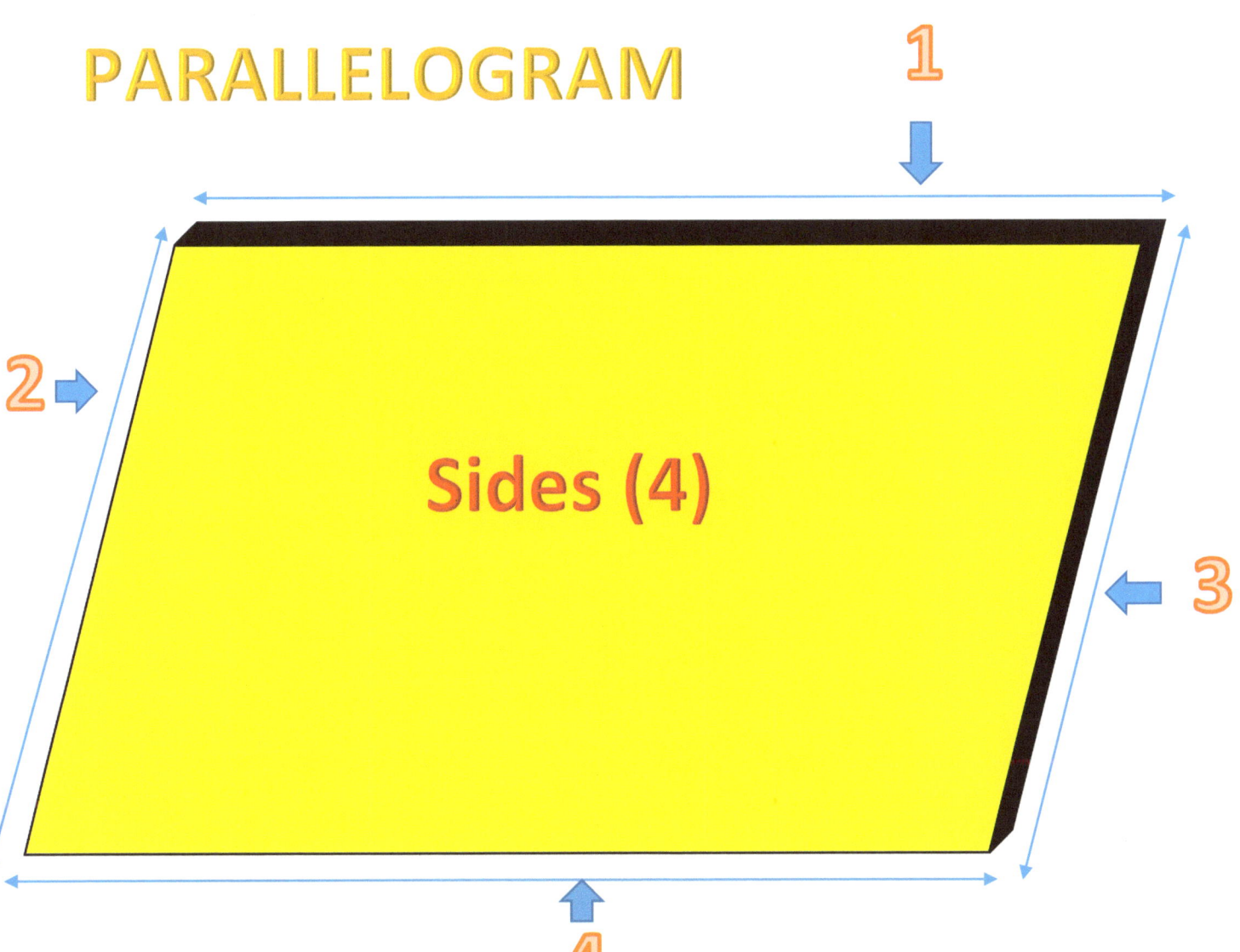

What gives a parallelogram its unique identity is that both pairs of opposite sides are parallel. In addition, pairs of opposite sides have equal length as well as opposite angles have equal measure

# TRAPEZOID

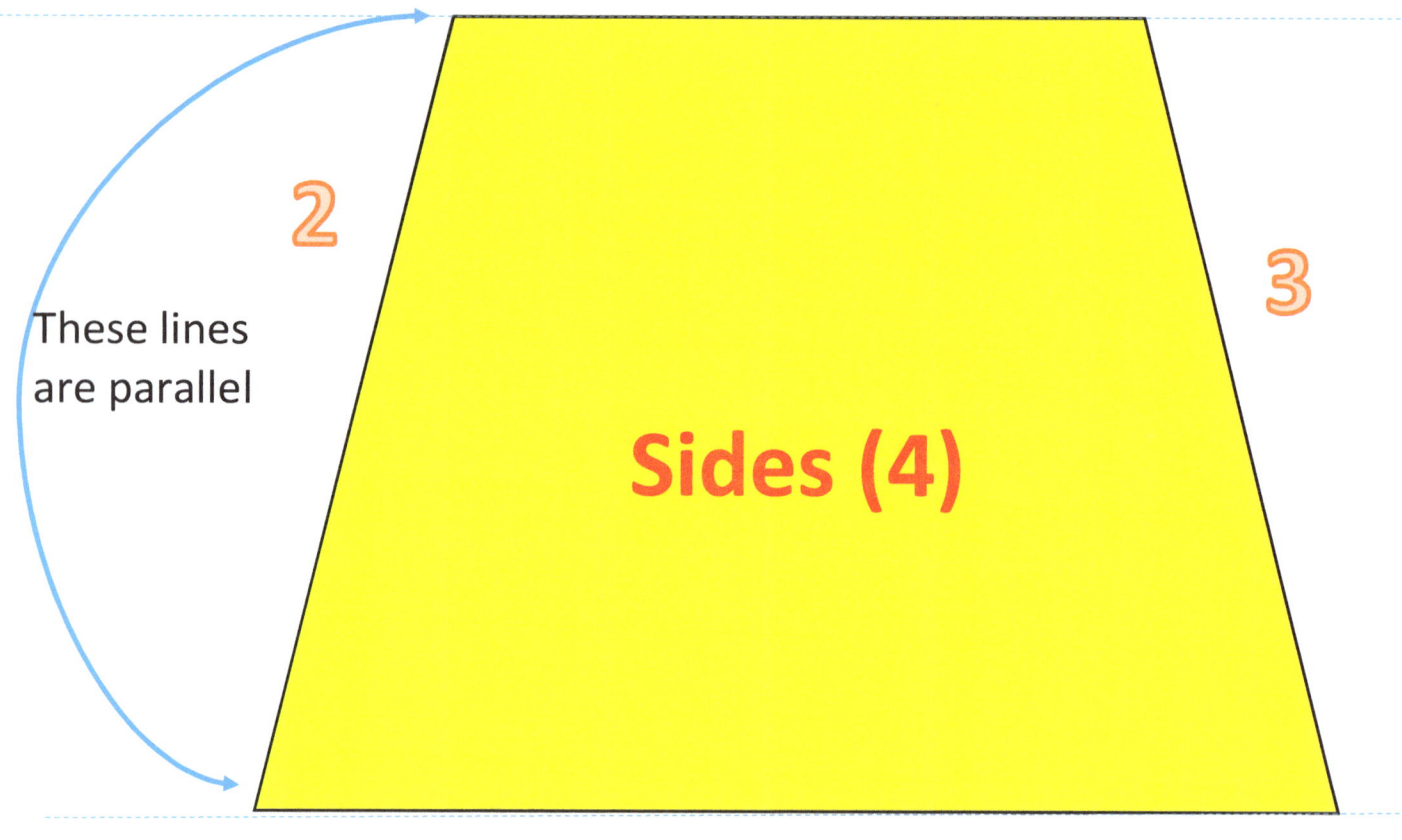

1. In a trapezoid one pair of opposite sides is parallel
2. These lines are parallel
3. 
4. The only rule a trapezoid must always abide by is that one pair of opposite sides must be parallel. The dashed lines represent parallel lines in which 2 parallel sides of quadrilateral lie.

Sides (4)

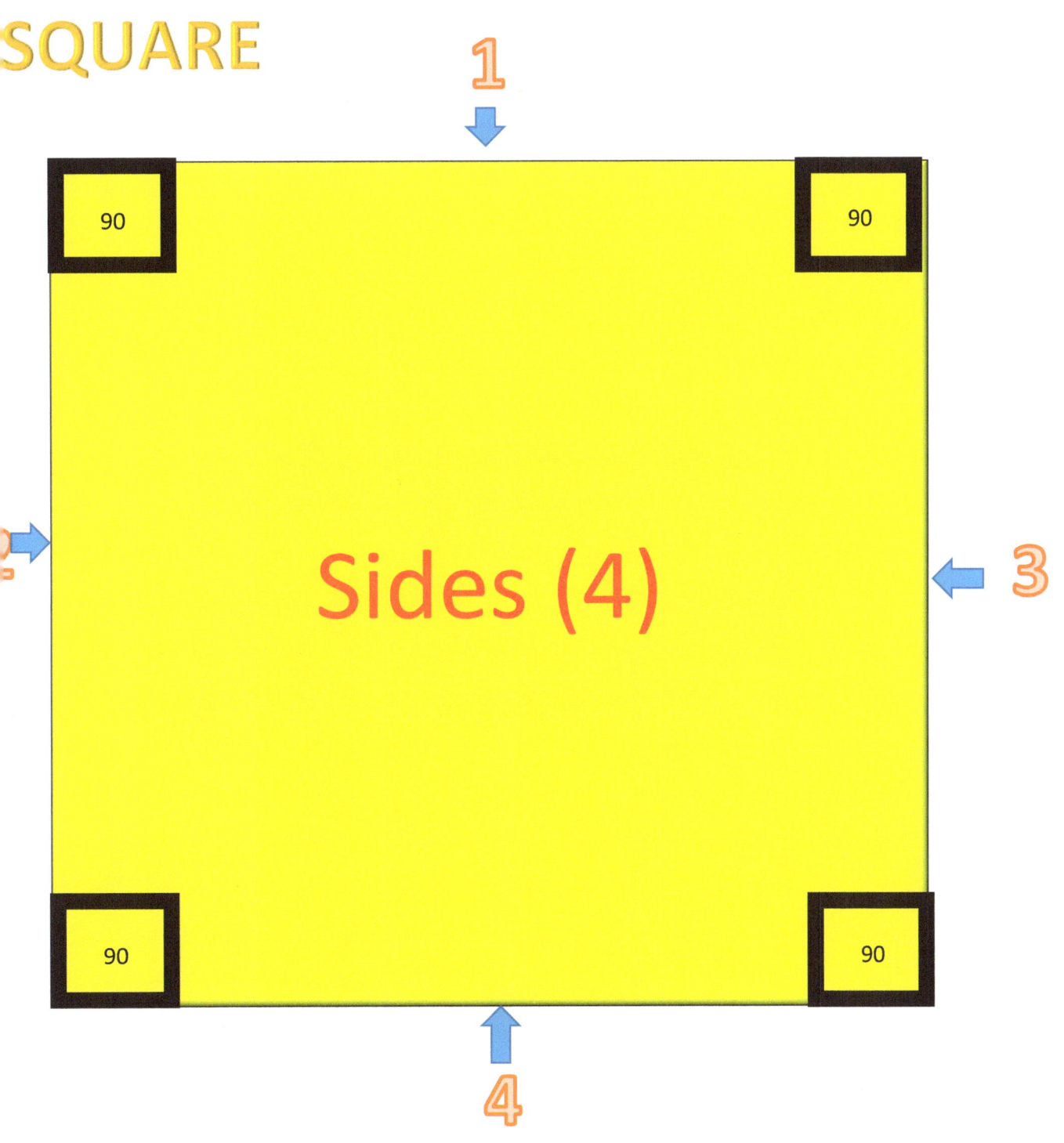

A SQUARE HAS FOUR SIDES THAT ARE ALL OF THE SAME LENGTH. FURTHERMORE, ALL THE INSIDE ANGLES ARE THE SAME, AND MEASURE 90 DEGREES. THE WIDTTH AND HEIGHT CANNOT BE ZERO.

# THE RING

# THE SPHERE

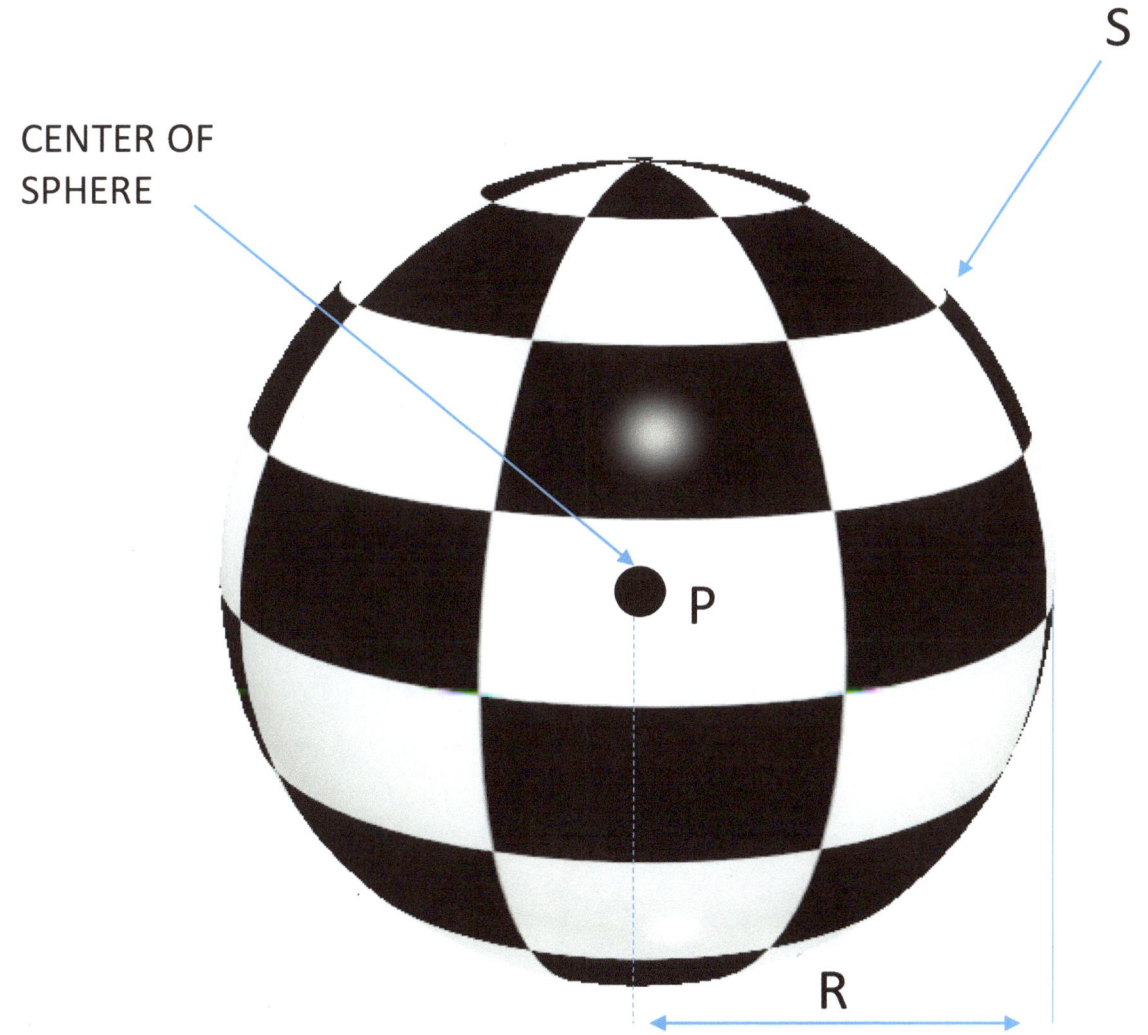

CONSIDER A SPECIFIC PIONT P IN 3D SPACE. THE SURFACE OF A SPHERE S CONSISTS OF THE SET OF ALL PIONTS AT A SPECIFIC DISTANCE OR RADIUS R FROM PIONT P. THE INTERIOR OF SPHERE S, INCLUDING THE SURFACE, CONSISTS OF THE SET OF ALL PIONTS WHOSE DISTANCE FROM PIONT P IS LESS THAN OR EQUAL TO R. NOT INCLUDING THE SURFACE, CONSISTS OF THE SET OF ALL PIONTS WHOSE DISTANCE FROM P IS LESS THAN R

# THE TORUS

A torus is a surface of revolution generated by revolving a circle in 3d space about an axis coplanar with the circle. A torus should not be confused with a solid torus which is formed by rotating a disc, rather

# RHOMBUS

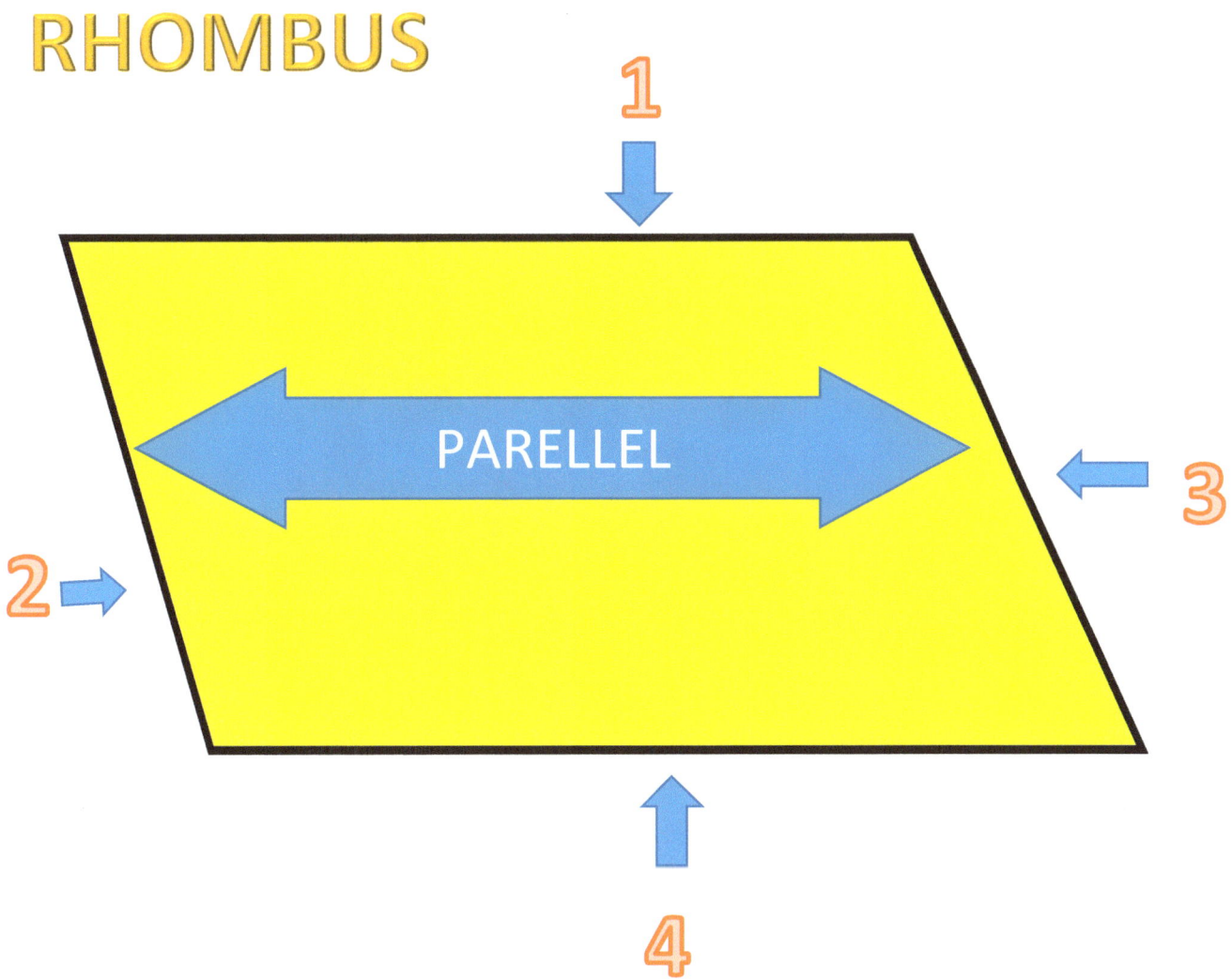

A RHOMBUS IS LIKE A SQUARE IN THAT ALL FOUR SIDES ARE THE SAME LENGTH. BU THE ANGLES DON'T ALL HAVE TO BE RIGHT ANGLES. ANOTHER PROPERTY OF THE RHOMBUS IS THE FACT THAT BOTH PAIRS OF OPPOSITE SIDES ARE PARALLEL.

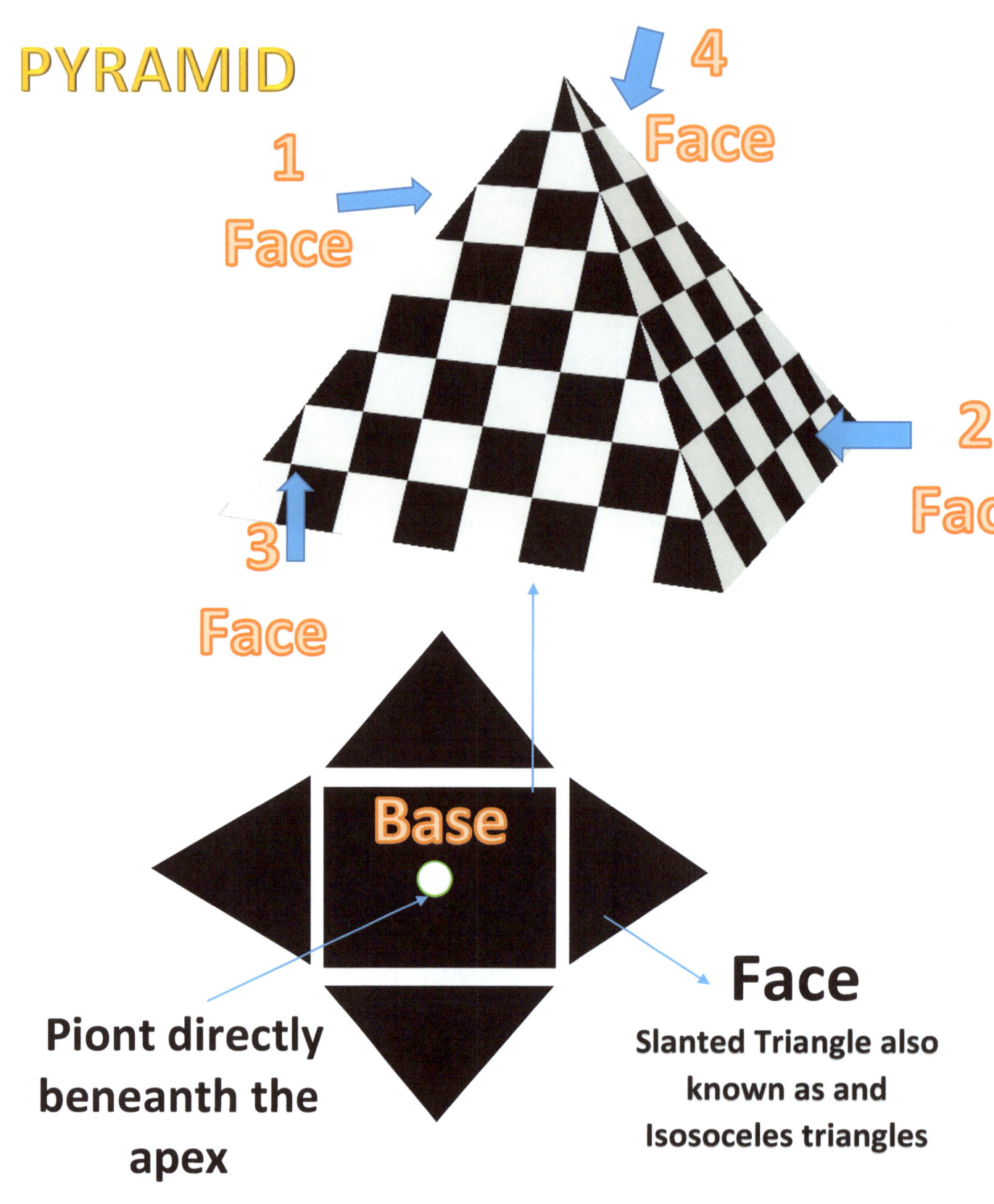

# Can you name theses shapes?

# QUIZ

1. HOW MANY SIDES DOES A HEXAGRAM HAVE?

2. WHAT IS A CUBE?

3. HOW MANY SIDE DOES A TRIANGLE HAVE?

4. WHAT IS THE ONLY RULE FOR TRAPEZOID?

5. WHAT ANGLES KIND YOU FIND IN A SQUARE?

6. NAME 2 SHAPES THAT CONTAIN PARALLEL LINES?

7. HOW MANY FACES CAN YOU FIND IN A PYRAMID?

8. WHAT SHAPE IDENTIFIES AS A HEXAHEDRON

# ANSWERS

1. A HEXAGRAM HAS 5 SIDES THE ROOT MEANING OF THE WORD HEX IS 5.

2. A CUBE is a three-dimensional solid object bounded by six square faces facets or sides with three meeting at each vertex. The cube also has 6 faces, 12 edges and 8 vertices

3. A TRIANGLE HAS 3 SIDES..

4. THE ONLY RULE FOR TRAPEZOID IS

5. 90 DEGREE ANGLES

6. TRAPEAZOID AND A PARALLELOGRAM

7. THERE ARE 4 FACES IN A PYRAMID

8. THE CUBE

www.ingramcontent.com/pod-product-compliance
Lightning Source LLC
Chambersburg PA
CBHW051839210526
45473CB00005B/1946